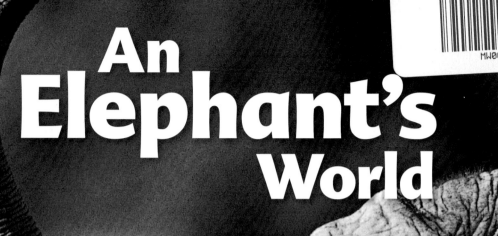

An Elephant's World

by Dennis Richard

Contents

Science Vocabulary

living
Living things are alive.

Elephants are **living** things.

A rock is a **nonliving** thing.

Food and water are **basic needs.**

shelter

A **shelter** is a safe place where a living thing can make its home and grow.

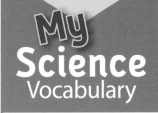

Elephants use trees for **shelter.**

basic needs
living
nonliving
nutrients
shelter

nutrients

Nutrients are parts of food and soil.

This baby elephant gets **nutrients** from drinking its mother's milk.

Living and Nonliving Things

This baby elephant walks with its mother. They are both **living** things.

A calf is a baby elephant.

living

Living things are alive.

The elephants walk over rocky ground.
Rocks and soil are **nonliving** things.

nonliving

Nonliving things are not alive and
never were.

How are living and nonliving things different? You can compare an elephant to a rock.

Living Thing: Elephant	Nonliving Thing: Rock
• An elephant must have air.	• A rock doesn't need air.
• An elephant must have food and water.	• A rock doesn't need food and water.
• An elephant moves on its own.	• A rock can't move on its own.
• An elephant can grow.	• A rock can't grow.

Basic Needs of Animals

Living things have **basic needs.** Food is a basic need.

basic needs

Basic needs are what living things must have to stay alive.

Nonliving things do not need food.

They do not have basic needs.

These rocks are in a part of Africa where elephants live. They are nonliving things.

Elephants must have food. A new calf gets milk from its mother. The milk has **nutrients.**

Nutrients help a calf grow.

nutrients

Nutrients are parts of food and soil.

Older elephants eat grass and trees. Most elephants eat for many hours each day.

Elephants eat all parts of the tree. Nutrients from leaves help elephants stay healthy.

Elephants also must have air and water.
They breathe through their trunks.

trunk —

An elephant's trunk has two nostrils for breathing.

Elephants drink water using their trunks, too.

Elephants suck water into their trunks. Then they put the water in their mouths to drink.

Elephants must have **shelter.** They stand under trees to get out of the sun.

Shade from trees protects elephants' skin.

shelter

A **shelter** is a safe place where a living thing can make its home and grow.

Elephants also must have space. They need space to roam, or walk.

Elephants walk long distances. They look for food and water.

A herd is a group of elephants. The herd moves slowly, so the calves can keep up.

In many herds, an older female elephant picks a place to stop to eat or drink.

Basic Needs of Plants

Elephants live in a grassland. Grasses and some trees grow in a grassland. These plants are living things, too.

Plants have basic needs. They must have air, water, light, nutrients, and space to stay alive.

Acacia trees are plants that grow in the grassland where elephants live.

The parts of the acacia tree work together to keep it alive.

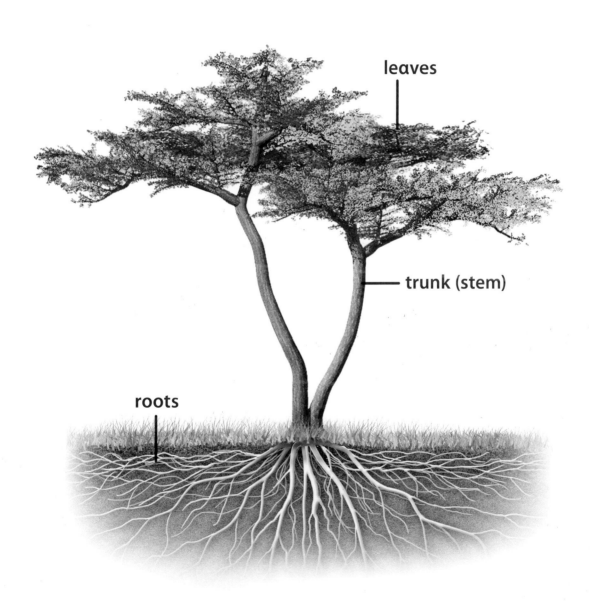

leaves

trunk (stem)

roots

All plants must have space to live and grow. Acacia trees must have space, too. Their branches grow wide. Their roots grow deep and can reach water below ground.

A grassland does not have much water.
But it is a good place for acacia trees
and elephants to live and grow.

Conclusion

Elephants and acacia trees are living things in a grassland. There are nonliving things in a grassland, too. The living things have basic needs. The nonliving things do not.

Think About the Big Ideas

1. How are living and nonliving things different?
2. What are the basic needs of acacia trees?
3. What are the basic needs of elephants?

Share and Compare

 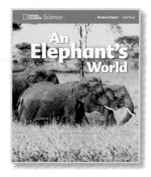 

Turn and Talk

Compare the living and nonliving things in your books. How are they different?

Read

Find a photo with a caption and read it to a classmate.

Write

Describe the basic needs of a living thing in your book. Share what you wrote with a classmate.

Draw

Show an animal getting something it must have to stay alive. Share your drawings with a classmate.

Meet Mireya Mayor

Scientists work together to solve problems. They try many ways to answer their questions.

Mireya Mayor helped discover a mouse lemur. She and her team wanted to catch it. They used a water bottle to make a trap, but the trap didn't work. The team finally found a way to catch the lemur. How do you think they did it?

Mireya Mayor holds a mouse lemur.

Index

Acknowledgments
Grateful acknowledgment is given to the authors, artists, photographers, museums, publishers, and agents for permission to reprint copyrighted material. Every effort has been made to secure the appropriate permission. If any omissions have been made or if corrections are required, please contact the Publisher.

Photographic Credits
Cover (bg) Polka Dot Images/Jupiterimages; Cvr Flap (t), 5, 10-11(bg, inset) Christian Musat/Shutterstock; Cvr Flap (c), 26-27 James Hager/Getty Images; Cvr Flap (b), 15 Neil Wigmore/Shutterstock; Title (bg) ImageState/Alamy Images; 2-3 Benoit Beauregard/iStockphoto; 4, 8-9 Yvette Cardozo/Alamy Images; 6, 11 (inset), 12, 28 Galen Rowell/Corbis; 7 (t), 18 Kondrachov Vladimir/Shutterstock; 7 (b), 14 Deborah Benbrook/iStockphoto; 13 George F. Mobley/National Geographic Image Collection; 16 Chris Johns/National Geographic Image Collection; 17 EcoPrint/Shutterstock; 19 Jack Hollingsworth Photography/PhotoDisc/Getty Images; 20-21 Beverly Joubert/National Geographic Image Collection; 22-23 Michael Nichols/National Geographic Image Collection; 24 Lockenes/Shutterstock; 30-31 Mark Thiessen/National Geographic Image Collection; Inside Back Cover (bg) Creatas/Jupiterimages.

Illustrator Credits
25 Paul Mirocha

Neither the Publisher nor the authors shall be liable for any damage that may be caused or sustained or result from conducting any of the activities in this publication without specifically following instructions, undertaking the activities without proper supervision, or failing to comply with the cautions contained herein.

Program Authors
Randy Bell, Ph.D., Associate Professor of Science Education, University of Virginia, Charlottesville, Virginia; Malcolm B. Butler, Ph.D., Associate Professor of Science Education, University of South Florida, St. Petersburg, Florida; Kathy Cabe Trundle, Ph.D., Associate Professor of Early Childhood Science Education, The Ohio State University, Columbus, Ohio; Nell K. Duke, Ed.D., Co-Director of the Literacy Achievement Research Center and Professor of Teacher Education and Educational Psychology, Michigan State University, East Lansing, Michigan; Judith Sweeney Lederman, Ph.D., Director of Teacher Education and Associate Professor of Science Education, Department of Mathematics and Science Education, Illinois Institute of Technology, Chicago, Illinois; David W. Moore, Ph.D., Professor of Education, College of Teacher Education and Leadership, Arizona State University, Tempe, Arizona

The National Geographic Society
John M. Fahey, Jr., President & Chief Executive Officer
Gilbert M. Grosvenor, Chairman of the Board

National Geographic School Publishing
Hampton-Brown
www.NGSP.com

Printed in the USA.
RR Donnelley, Wetmore, TX

ISBN: 978-0-7362-7578-1

11 12 13 14 15 16 17

10 9 8 7 6 5 4 3